James Watt was one of the pioneers of the science of mechanical engineering. His revolutionary ideas about steam engines helped to ensure that Britain was the first truly industrial nation. His ability to master a wide range of subjects brought him praise and admiration from scholars and engineers alike. He is best remembered as the developer of the steam engine, but he was also a skilled craftsman, a civil engineer, designer and surveyor, and taught himself a number of foreign languages. Andrew Nahum presents a vivid picture of this rather serious scientist, as a young artisan; a skilled engineer, and an eminent figure. The book is illustrated with over 40 black and white drawings and photographs. A glossary, reading list, and an index are also included.

Andrew Nahum is a graduate from Edinburgh University. He is at present a Curator at the Science Museum in London.

Pioneers of Science and Discovery

James Watt

and the Power of Steam

Andrew Nahum

Other Books in this Series

First published in 1981 by
Wayland Publishers Ltd
49 Lansdowne Place, Hove
East Sussex BN3 1HF, England

© Copyright 1981 Wayland Publishers Ltd
ISBN 0 85340 826 2

Phototypeset by Computacomp (UK) Ltd
Fort William, Scotland
Printed in Italy by G. Canale & C.S.p.A., Turin

Contents

Below James Watt Tavern. The site of the house where James Watt was born.

1 *Introduction*

James Watt was one of the first British engineers to become a famous national figure. Before his time mechanical subjects had little prestige. By the time he died, engineering had developed into a profession, with many spectacular successes. Through his steam engine developments Watt came to be regarded as an 'industrial hero' and 'one of the real benefactors of the world'. His importance in engineering was even compared to that of Shakespeare in poetry.

During Watt's life, enormous changes occurred in society. Industry was growing at a tremendous rate – as an indication, iron production doubled every eight years between 1780 and the early 1800s. Britain was becoming the most important manufacturing nation on earth and the mood of the age was one of optimism. Much of the initial surge in manufacturing was powered by water, or even horses. However steam power was becoming increasingly important and in the popular imagination, was seen as one of the main agents of the new prosperity – the tireless power source that was freeing man from back-breaking physical labour. Watt was viewed as the man who had made this possible.

Watt was honoured by both scientists and men of letters. They respected him because they shared his view of himself – as a thinker who had brought about practical steam power by experiment and theoretical study. Some historians have claimed that progress in eighteenth-century technology sprang from the scientific discoveries of the period. Others argue that much science

the Cylinder at Kinneil is 18 in
Diameter its area 254½ Sq inches
the pump is 18½ Inch diameter
& 25 feet high ...
... area 269 in
Contents 46 ⅔ Cub feet
weight. 2893 pounds
being 10 ⅔ to the Inch —
on the 16th of Feb) 10th of Salt.
disolved in water was beginning
to be poured daily on ⅛ Cub y
Cakes in Cellar at Kinneil
on the 10th of March one half of the
brine was sett apart & the other
half diluted with it, own bulk.
of water & ... process —

the pressure on Cylinder at
Kinneil is 3564 pounds at 14
to the Inch. If the steam be = 3
it amounts to 750 lb = small to
4314 pounds which Surpasses
the weight of ▽ 1421 pounds

for this Condenser the part between
the two pumps is supposed to retain
the heat given it by the steam —
& here by cause the water to boil
lyes between the valves which hind
the Condensation for the engine on
made about 6 strokes a minute & worki
the Condenser faster did not accelerate th
process the condn was not hindered b
air for very little was put out nor di
it appear that much was disengaged
from the ▽ entering the Condr & ever

Above Notes taken from James
Watt's Journal. Written in 1770 and
describing the engine at Kinneil.

(particularly the study of heat) developed from the
insights of practical craftsmen who had to make
steam engines work. In fact science and
craftsmanship were both combined in James
Watt's personality. He was certainly a craftsman,
and was noted for his skill with tools throughout
his life. However he saw himself as a scientist (or
'natural philosopher') who had solved mechanical
problems by scientific method and observation of
nature.

A great deal of Watt's correspondence survives,
and tells us much about his life and the progress of
his inventions. His letters sometimes reveal him as
excessively serious. For example, he writes that he

Above The house at Heathfield, near Birmingham where Watt lived for many years.

will engage John Southern as a draftsman, 'provided he gives a bond to give up music, otherwise I am sure he will do no good, it being the source of idleness'. On another occasion, advising a friend on the education of her son, he recommends that he be 'thoroughly taught arithmetic and algebra ... Cultivate his taste for drawing in India ink but go no further lest it degenerate into painting'. Watt also seems often to have been self pitying and beset by doubts. No one, however, could have wished for more loyal friends – friends drawn from among the greatest men of his day. From his beginnings in Glasgow he enjoyed support and encouragement from people who recognized his potential. Certainly Watt must have had an engaging side to his character. Touches of wry humour that peep through from time to time in his letters hint at this.

Late in life, Watt built himself a pleasant house, Heathfield, near Birmingham. A small garret room under the roof was converted into a workshop and he spent time there after his retirement, working on various projects. In 1819, Watt died and his workshop was locked. It lay undisturbed for about forty years. When it was re-opened, dust lay two inches thick on the floor and the first visitors made footprints as if walking in soft snow. The workshop became a famous spot for those interested in industrial history. One of Watt's biographers, J. P. Muirhead, described it in terms that seem to us now almost religious – 'The classical garret and all its mysterious contents … have ever since been preserved in the same order as when the hand and "eye of the master" were last withdrawn from them, and he crossed the threshold never to return to his work on earth … all things there seemed still to breathe of the spirit that once gave them life and energy … no profane hand had been permitted to violate the sanctities of that magical retreat.'

In 1924 Heathfield House was demolished. The room itself was reconstructed in the Science Museum, London, using the woodwork and fittings from the old house. All the contents were re-arranged in their original places in the room. The workshop is an intriguing place. Almost all the phases of Watt's life are represented there. There are plates for printing barometer faces and many other relics of his first career as an instrument maker in Glasgow: a surveying quadrant of his own design, partly built, with

Above A view of Watt's garret workshop showing his last experiment – the sculpture copying machine.

Below A close-up view of a barometer made by James Watt.

pieces wrapped in paper torn from a notebook of his own surveying projects – a reminder of another of Watt's trades. There are unfinished parts of musical instruments too – these were a sideline of his Glasgow shop. Perhaps the most interesting item is an experimental model steam engine, incorporating Watt's famous improvement – the separate condenser.

After his celebrated career as a steam engineer, Watt worked in the room on machines to copy sculpture. The principle is that a feeler passes over the original, and controls the movement of a rotating cutter. The cutter should cut a stone blank to faithfully reproduce the original. To make a machine that could copy complex shapes like human figures was a very difficult task. However Watt had achieved fair success when he died. (Today the 'copy mill' is an established industrial tool.) One of his machines was designed for making smaller copies. In one of his last letters he wrote modestly, 'If I live I ... hope to be able to produce a reduced copy of Chantrey's bust of myself, fit for a chimney piece, as I do not think myself of importance enough to fill up so much of my friends' houses as the original bust does'. A full-size copy of Sir Francis Chantrey's bust of Watt stands on a bench in the workshop.

Today we can view the room through a window where one wall once stood. Everything is much as it was when Samuel Smiles, great chronicler of the Industrial Revolution, visited Heathfield and wrote, 'The piece of iron he was last employed in turning lay on the lathe. The ashes of the last fire were in the grate, the last bit of coal was in the scuttle. ... The frying pan in which he cooked his meal was hanging by its accustomed nail. ... Near at hand is the sculpture machine, on which he continued working to the last.'

Below Another view of Watt's 'copy-mill' or sculpture machine.

2 The Young Instrument Maker

James Watt was born on the 19th of January, 1736 at Greenock on the Clyde. The town was then growing in importance as a port and Watt's father prospered with it. He had been trained as a carpenter or 'wright'. However he seems to have developed various business interests including house and ship building and a ship's chandlery. Later he became a general merchant and part-owner of some ships. He must have been a respected and trusted man in the community for in 1755 he was Treasurer of the Burgh.

James Watt was a weak child and his mother nursed him through long bouts of illness. Although he recovered and lived to be 83, he always lived in great fear of illness and often complained of poor health. When Watt was 31 his great friend, Dr Erasmus Darwin, wrote to him, 'I first hope you are well and less hypochondriacal.'

When he was 10 or 11 years old Watt was judged strong enough to go to school. After his solitary childhood he seems to have hated the noise and rough and tumble of school life and was thought to be dull. However at 13 he moved to the Grammar School in Wee-kirk Street and soon shone at mathematics. At the same time he had a workbench and forge in his father's shop, where he developed his abilities as a craftsman. At 18 Watt decided to become a scientific instrument maker. In his father's chandlery he must have handled various types of ships' instruments. No doubt he felt that in this career he could combine

Left Glasgow University buildings. This view was probably the same when James Watt had his business within the University precincts.

his practical skill with his interest in mathematics. Watt travelled to nearby Glasgow to train.

At that time the instrument trade in Scotland was too small to allow much specialization. Like Watt himself many of the instrument makers were highly versatile men. There was then no real profession of 'engineer' and Scottish instrument makers were consulted on very varied tasks. They were capable of some of the most accurate work available and could often combine this with knowledge of mathematics and contemporary science. For example, when John Bond designed a novel harpoon gun for an Edinburgh whaling company he turned to an instrument maker, Thomas Short, to make it. Watt's first master was said to have made drawing instruments, fishing tackle and cutlery, as well as being a tinsmith and 'a repairer of fiddles and a tuner of spinets, a useful man at almost everything'.

Watt was related to George Muirhead, Professor of Humanities at Glasgow University and through him was introduced to other scientific men. At this time the Scottish universities were acknowledged to be the best in Europe. They were also far more 'open' and democratic than their English counterparts. Perhaps this helped Watt meet eminent men, for he did not have a normal education and was learning a 'trade'. However he did have an acutely enquiring mind, coupled with a wide knowledge gained from his eager reading. Certainly the professors were impressed by him. One of them, Robert Dick, taught Natural Philosophy, which was Watt's particular interest. (Natural Philosophy was broadly equivalent to what today we call Physics.) Dick encouraged Watt to go to London to train, for the instrument makers there were producing work of ever

Above The Press-gangs were used to force men to serve in the Navy. Apprentices to the City guilds were exempt but any other individual risked being kidnapped if they walked through the streets at night.

increasing accuracy and were virtually unrivalled in the world.

In June 1755, financed by his father, Watt set out. He travelled on horseback, with a companion, and the journey to London took them two weeks. Watt had a letter of introduction from Professor Dick to James Short. Short too was a Scotsman, established in the Strand and famous for his telescopes. James Short was unable to offer Watt a job. After one month James Watt wrote to his father 'I have not yet got a master, they all make some objection or other'. The problem was the power of the trade guilds. Watt had not served an apprenticeship and therefore could not officially be employed. Neither could he become an apprentice since he was too old. Eventually John Morgan of Cornhill agreed to teach Watt for one year at a fee of twenty guineas. Watt wrote 'if it had not been for Mr Short I could not have got a man in London that would … teach me, as I now find there are not above five or six that could have taught me all I wanted'. Watt set himself to learn in one year what normally took three or four. Work went on till nine at night, while he seldom went out for fear of the press gangs who were seizing men for the navy. The gangs were active, on account of the Seven Years' War with France. Although they were not supposed to take apprentices or tradesmen in the City of London, Watt knew that his unofficial status would give him no protection if he were taken.

After a year of this application Watt wrote, 'I think I shall be able to get my bread anywhere'. He had lived on eight shillings a week, for his father could ill afford to keep him and by now he was suffering from 'a racking cough, a gnawing pain in the back and weariness all over the body'.

Watt returned to Scotland and spent some time at home in Greenock. He then moved to Glasgow where his first job was to clean and repair a very fine collection of astronomical instruments just brought back from Jamaica. In 1757, he was provided with a shop in the University buildings and allowed to call himself 'Mathematical Instrument Maker to the University'.

Watt's shop became a meeting place for a circle of men who shared his curiosity about Natural Philosophy. One of his early visitors was Dr Joseph Black, the most eminent chemist of this period. Another was John Robison, then a student, but destined to succeed Black to the Chair of

Above The Quadrangle at Glasgow University. It was here that James Watt founded his instrument making business. The shop also became a popular meeting place for many of the professors and students at the University.

16

Chemistry at Edinburgh. The friendship of these three men was to be lifelong. Robison recalled his first meeting with Watt in these words, 'I saw a workman and expected no more – but was surprised to find a philosopher, as young as myself, and always ready to instruct me'. That the more formally trained academics were full of respect for Watt's ceaseless curiosity and self-taught knowledge shows from the letters and memoirs that survive. Once Watt was commissioned to build an organ. Robison recalled that although Watt did not know one musical note from another, 'We imagined Mr Watt could do anything'.

The organ was duly built, but in addition Watt led his friend into a deep theoretical study of musical harmony. Robison's judgement has become famous, 'Every thing became to him a Subject of new and serious Study – Every thing became Science in his hands'.

Over the next few years Watt seems to have been fairly prosperous. He acquired a partner and moved shop. In addition to mathematical and 'philosophical' instruments he also made guitars, harps, flutes and bagpipes.

In 1763 an event occurred that was to be crucial in Watt's life. He was asked to repair a model of a

Newcomen steam engine for the University Natural Philosophy class. Newcomen had invented his 'atmospheric' engine fifty years before. During the eighteenth century there was a growing demand for coal and metals. The Newcomen engine was widely used throughout the country as the only practical way of pumping out the water that always threatens to flood deep mines.

The demonstration model was supposedly to perfect scale but it would only run for a few strokes. Watt set out to find why. He was not content, however, just to make it work. He wanted to understand the principles involved. He started a series of experiments on the properties of steam and made accurate measurements. For some of these experiments Watt used an early form of pressure cooker as his boiler.

At this time the scientific world had rather vague ideas about the properties of heat. Phenomena like radiation and conduction were being investigated. So too was the idea that different materials could absorb different amounts of heat. However the distinction between temperature (a state) and heat (a quantity) was far from clear. In the course of his experiments Watt discovered a puzzling thing. Steam appeared to contain much more heat than an equivalent weight of water at the same temperature. He had been investigating condensation by passing steam from a pipe into a vessel containing cold water. (Condensation played an important part in the Newcomen engine.) As the steam condensed, the water level rose, but the volume had only increased by 1/6th before all the water was at boiling point and could condense no more steam. Nothing Watt had read or heard made him expect this. On the accepted law of mixtures the small quantity of condensed water should have had a

Above A portrait of Dr Joseph Black, Professor of Chemistry at Edinburgh University. He had a life-long admiration for Watt.

proportionally small effect on raising the temperature. Watt later recalled, 'Being struck with this remarkable fact and not understanding the reason for it, I mentioned it to my friend Doctor Black, who then explained to me his doctrine of latent heat, which he had taught for some time'. Black confirmed that when a quantity of water over a fire reaches boiling point it continues to absorb heat. The heat however goes to change the state of water into steam, without a further rise of temperature. This 'latent' heat is returned on condensation and had produced the effect that Watt had noticed. He calculated the latent heat of steam and got a figure of 534 – very close to the modern figure.

Watt had given himself a better scientific insight into the properties of heat and steam than any engineer of his day. He measured how steam pressure increased with temperature and expressed his results as a graph – then a very unusual technique. He was now almost uniquely qualified to devise the revolutionary improvements to the steam engine that were to bring him fame and respect.

3 *The Fascination of Steam*

Left James Watt's close study of the scale model of Newcomen's Engine marked the beginning of his interest in steam engineering.

The Newcomen engine which Watt began to study had become an invaluable aid in mining since the first known example was built in 1712. Thomas Newcomen had been an ironmonger in Devon. However the steps that led to his engine appear to be a mystery. It seems that he visited the Cornish tin mines to supply tools and realized the difficulties of pumping them free of water. Fifteeen years of development produced his atmospheric engine but unfortunately we know far too little about him and his experiments. Not even a single portrait of him survives. Discoveries about atmospheric pressure and the power that it could exert were known in the small scientific community of the time. They were published, for example in the Philosophical Transactions of the famous Royal Society of London – the most important scientific society of its day. However these interests were mainly confined to a few major centres, and scholars have been unable to learn whether Newcomen could have heard about them in the West Country, which was a fairly isolated area at that time.

Many heartbreaking disappointments must have occurred during the long years of development. However we have no clue as to how these experiments were financed. History has been unkind to Newcomen. He was undoubtedly a man of tremendous perseverance and probably of genius. Some accounts though have called his engine 'little more than a combination of known parts' or even, merely, the result of a series of 'happy accidents'. Samuel Smiles, the famous

Victorian chronicler of the Industrial Revolution, but not an engineer, described the working of the Newcomen engine as 'a clumsy and apparently very painful process, accompanied by an extraordinary amount of wheezing, sighing, creaking and bumping. When the pump descended, there was heard a plunge, a heavy sigh, and a loud bump; then as it rose, and the sucker began to act, there was heard a creak, a wheeze, another bump, and then a rush of water as it lifted and poured out'.

In fact though, Newcomen had produced a very effective device and there had never been anything like it before. For more than 60 years it remained the only practical method of keeping deep mines free of water. Its importance to the eighteenth-century economy can be judged by the speed with which the invention spread. In 1722, only 10 years after the first recorded example was built at a mine in Staffordshire, Newcomen engines were erected at Königsberg (a town in Czechoslovakia). By 1775 the engines were widespread in mining districts throughout Britain. For example there were over 60 working in Cornwall alone, and about 100 at collieries in the Newcastle area.

The principle of the Newcomen engine was to use the power of atmospheric pressure to lift a quantity of water from the mine at every stroke. One end of the rocking beam was connected by rods to the pump deep in the mine; the other to a piston working in a steam cylinder. When the piston rose, steam from the boiler was admitted to the cylinder under the piston. Then – one of Newcomen's notable discoveries – water was sprayed into the cylinder to condense the hot steam. Since the condensed steam only took up a fraction of the space formerly occupied by the steam, a partial vacuum was produced in the

Above Von Guericke's experiment with two hemispheres to prove the power of atmospheric pressure. The spheres fitted together tightly, the air was then pumped out to form a vacuum. It was then impossible for two teams of eight horses to pull the spheres apart.

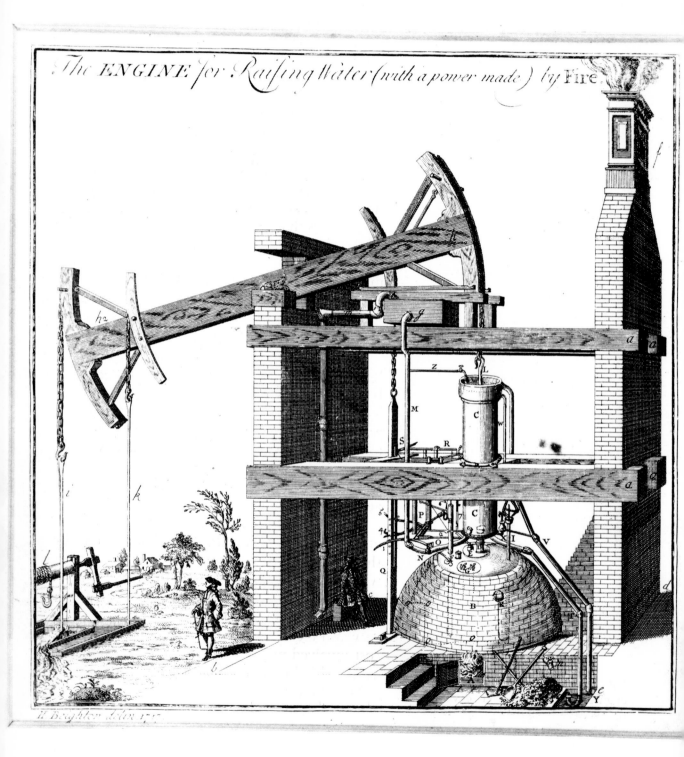

The ENGINE for Raising Water (with a power made) by Fire.

T. Beighton delin. 1717

24

cylinder and atmospheric pressure forced down the piston. The beam thus rocked down, pulling up on the pump rods and raising water from the mine.

Newcomen must have been beset by difficulties, particularly the problem of sealing his piston in the cylinder, for contemporary techniques could not produce an accurate cylindrical bore. He solved this by using a flexible leather washer on his piston, and keeping a layer of water above it. Like many successful inventions, the beauty of the engine was that it suited the manufacturing skills of the time. Many ingenious schemes had failed because the existing technology had been too inadequate to put them into practice. James Watt himself was to endure many difficulties before he was able to incorporate his theoretical insights into a practical full-size engine.

Newcomen's engine is often called an 'atmospheric' rather than a 'steam' engine. This is because in this type of engine the piston is not driven by steam pressure. Instead the steam is used to produce a partial vacuum below the piston and allow normal air pressure to drive it down. The advantage of this system in the eighteenth century was that it used steam at only slightly above atmospheric pressure.

The blacksmithing art of the period was well able to produce boilers for this. In later years, after the age of the Newcomen and the Watt engines which followed them, the power of steam under pressure was exploited. This was done in the search for greater efficiency. However it required new techniques of boiler construction, and the cost involved many disastrous boiler experiments during development.

The Newcomen engine that the 'insatiably curious' James Watt came to study was then an

Left An engraving of a Newcomen engine. These engines were developed to pump water from the Cornish mines.

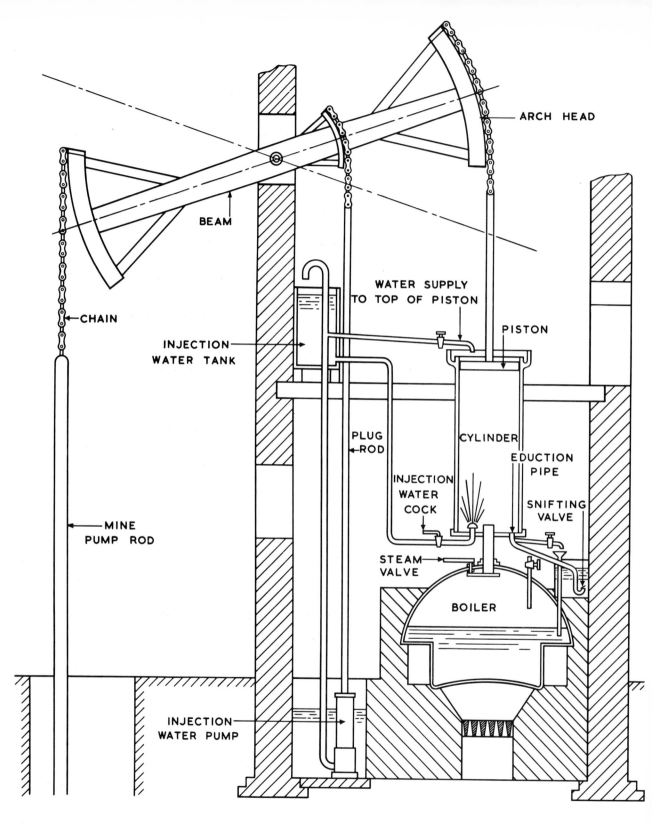

ARCH HEAD

BEAM

CHAIN

INJECTION
WATER TANK

PLUG
ROD

MINE
PUMP ROD

INJECTION
WATER
COCK

STEAM
VALVE

WATER SUPPLY
TO TOP OF PISTON

PISTON

CYLINDER

EDUCTION
PIPE

SNIFTING
VALVE

BOILER

INJECTION
WATER PUMP

26

almost commonplace industrial device. However it had received little benefit from study by educated or scientifically-minded people. Samuel Smiles remarked 'The educated classes of the last century (i.e. the eighteenth century) regarded with contempt mechanical men and mechanical subjects ... engineering was thought unscientific and ungenteel'. James Watt, with his interest in Natural Philosophy, was a new kind of student of these problems.

The chief defect of the Newcomen engine was its high consumption of coal. This was not so much a problem when it enabled coal mine owners to win coal from deeper levels than before. However it led to high costs in metal-mining areas like Cornwall which relied on imported fuel for pumping. The University's model that Watt had been asked to repair suffered from the same defect – a remarkable appetite for steam. Perhaps if the boiler of the model had been adequate to keep up with the engine's requirements Watt would not have been led to question the reason for its inefficiency. His experiments made him realize that the reason lay in the alternate heating and cooling of the cylinder. As steam entered the cylinder for each cycle, much of it was used to heat the metal walls to nearly boiling point again, after the previous condensing stroke. In fact the quantity of steam the model used at each stroke was several times the volume of the cylinder.

Watt struggled with the contradictory need to maintain the cylinder at high temperature, but still to condense steam with cold water to produce the vacuum for the working stroke. His mind must have turned over the problems of the steam engine endlessly. He related later 'It was in the Green of Glasgow. I had gone to take a walk on a fine Sabbath afternoon. ... I was thinking upon

Left Diagram which shows the working of a Newcomen engine.

Above Watt's house in Delftfield Lane, Glasgow.

the engine at the time'. At this moment he realized that as steam was a gas it could be condensed in a separate chamber connected to the working cylinder of the engine. Steam would flow into this chamber until condensation was complete. In Watt's words 'as steam was an elastic body it would rush into ... a vessel and might there be condensed without cooling the cylinder'. In this way the working cycle of the engine could carry on without the wasteful alternate heating and cooling of the cylinder.

There is a story that out of respect for the Lord's Day, Watt waited until Monday to build a model for the experiment he was bursting to try. Crudely made in tinplate, it proved that his thinking was correct. It can still be seen today in the Science

Above This is the original model of the separate condenser engine made by Watt in 1765.

Museum, London. It shows signs of hasty improvisation, unlike Watt's usual careful craftsmanship and perhaps shows his eagerness to test the idea.

Watt realized that his scheme could be the basis of a vastly improved engine and it has been called the greatest single improvement in the history of the steam engine. However it was to take many years and Watt was to suffer many disappointments before his theoretical insights were transformed into a practical machine. Watt was to write of these struggles, 'I have mett with many disappointments. I must have sunk under the burthen of them if I had not been supported by the friendship of Doctor Roebuck. I have now brought the engine near a conclusion, yett I am not nearer the rest I wish for than I was 4 years ago. ... Of all things in life there is nothing more foolish than inventing'.

Dr Roebuck was one of the most important figures in Scottish industrial history. Qualified in medicine, his interest in applied chemistry brought him from Birmingham to Scotland where he helped found a plant for making sulphuric acid outside Edinburgh. At this time Scotland had no iron industry, so Roebuck and his partners formed the famous Carron Iron Works on the banks of the River Carron. Production started in 1760, and 1,524 tonnes of iron were smelted in that year. Watt's friend, Doctor Joseph Black knew Roebuck through their interest in chemistry, and arranged an introduction, hoping that both might benefit. Roebuck eventually agreed to acquire a two-thirds share in Watt's steam engine improvements. For this he payed off a debt of £1,200 resulting from the death of Watt's partner in his instrument making business. In addition he agreed to bear the costs of patenting Watt's new ideas.

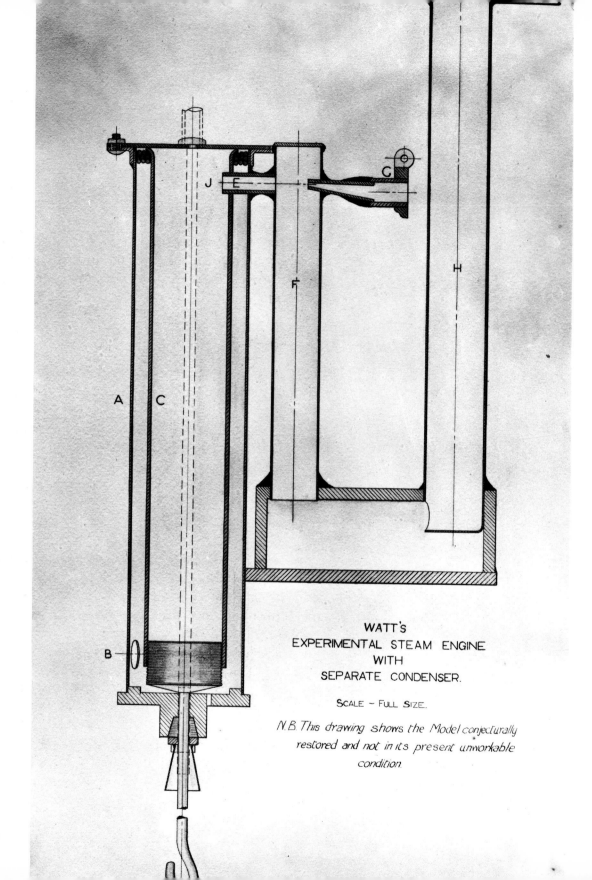

WATT'S
EXPERIMENTAL STEAM ENGINE
WITH
SEPARATE CONDENSER.

SCALE ~ FULL SIZE.

N.B. This drawing shows the Model conjecturally
restored and not in its present unworkable
condition.

4 Building an Engine

Watt was quite aware that he lacked experience of full-size engine building. In fact it seems that at the time of his steam experiments he had never seen a working Newcomen engine. His partnership with Roebuck started in about 1769 and, in order to learn more of full-scale practice, he helped build a Newcomen engine at the Carron Iron Works. Watt later recalled that he had built 'some very indifferent common engines' during this period.

The contrast between the accuracy of mathematical instrument making and the smithwork of the period was very frustrating to Watt. Many years later, he recalled: 'When I began first to construct my engine I found the workmen or Engineers accustomed to the erection of former engines so opinionated and obstinate that I had to discontinue employing them, and not only form my own Engines but also my own Engineers'. We know that he was disappointed with an experimental cylinder for his engine, 'the best Carron could make'. However it was the skills of the instrument makers that eventually formed the foundation for the new profession of mechanical engineering.

The first full-size Watt engine was built at Kinneil, Dr Roebuck's house, about 20 miles east of Edinburgh. During this period Watt wrote to friends 'I can think of nothing else but this engine', and also 'What I knew about the steam engine before ... was but a trifle to what I know now'. Watt laboured at Kinneil with the problems of his new engine. In a sense he had designed a device that contemporary workmanship was not

Left A diagrammatic representation of the separate condenser engine pictured on the previous page.

yet ready to build. A major part of his problems
came from the difficulty of sealing the piston in
the bore. Newcomen engines had used a leather
flap or a compressed ring made of hemp. A small
supply of water above the piston had helped keep
the seal steam-tight and compensate for the
inaccuracies of contemporary cylinder bores.

Watt had realized that the admission of cold air
above the piston was another cause of wasted heat.
He resolved to close the upper part of the cylinder
and allow boiler steam to drive the piston down.
The modest steam pressure used in Watt's boiler
gave some increase in power over atmospheric
pressure. His main object, though was, to improve
thermal efficiency and keep the cylinder as hot as
possible at all times. Unfortunately the water seal

Above Watt's rotative engine.

could no longer be used in a steam cylinder and Watt experimented with many combinations of packing and oils to keep the piston steam-tight in the bore.

In fact progress was slower than Watt and his friends had hoped. He still maintained his instrument making business where he produced guitars, violins and flutes in addition to scientific instruments. Dr Roebuck too was distracted by many schemes and does not seem to have provided the finance or encouragement needed.

At this time Watt started working as a civil engineer, mainly surveying routes in Scotland for various canal proposals. A large proportion of an instrument maker's business was in surveying equipment and Watt must have acquired a good understanding of their use. At this time Britain was entering the great age of canal building. Manufacture and commerce were increasing rapidly. Transport of raw materials was far more economical by water than by horse drawn carts over poor roads. For several years Watt was engaged on these schemes. He also designed a bridge over the River Clyde at Hamilton and the docks at Port Glasgow.

In 1767 Watt finished surveying a canal route to link the Forth and Clyde rivers. He visited London to try and gain Parliamentary approval for the proposals. He did not succeed. Perhaps for this reason he wrote to his wife about the House of Commons 'I never saw so many wrong-headed people on all sides gathered together. ... I believe the Deevil has possesion of them'. The next year Watt visited London again, to take out a Patent for his steam engine improvements. This was entitled '... a Method of Lessening the Consumption of Steam and Fuel in Fire Engines'. (The term, fire engine at that time implied a device that drew its power from fire – not a device for putting out fires! 'Steam engine' became the generally accepted term during Watt's lifetime.)

Watt returned by way of Birmingham and there he met Matthew Boulton. This meeting was to change the course of Watt's life and was of vital importance in the development of steam power.

Matthew Boulton was one of the major figures of the Industrial Revolution. His father had been a manufacturer of ornamental metal goods. However Boulton extended the business and in 1764 built a magnificent 'manufactory' at Soho, about two miles north of the city centre.

Birmingham had become famous in the eighteenth century for the manufacture of metal goods. This success was partly due to the absence

Above Port Glasgow. The original design for these docks was completed by James Watt.

of trade guilds – the system that had restricted Watt himself when he went to learn his trade in London. In Birmingham, craftsmen and small workshops flourished and produced an astonishing range of products. It had also become a place where religious dissenters could set up in business. Many of these were middle-class manufacturing or business people and the town

Left A view of the city of Birmingham as it appeared in the late 18th century. From this period onwards the town developed into a great industrial and commercial centre.

Right A portrait of Matthew Boulton. This man was a commercial genius and it was partly due to his foresight that the town of Birmingham developed into one of the major industrial centres of Europe.

became known for industries with a high degree of skill. The population of Birmingham doubled between 1770 and 1800, showing its commercial success. However the town also had a reputation – perhaps undeserved – for poor quality. The term 'Brummagem made' implied cheap and shoddy articles.

Boulton though was determined to make only products of good quality and design. In fact this was one of the ruling principles of his life. One of his innovations was to gather many different trades under one roof. His 'model' manufactory employed over 600 skilled craftsmen. It was one of the sights to see of the 'modern' age and drew a continual stream of visitors from all over Europe.

Boulton certainly realized that his system made it possible to make articles of consistent high quality at reduced cost. His methods anticipated the powerful modern techniques of mass-production. Among many things, the manufactory produced inlaid buttons, buckles, and tableware. Many pieces made at the Soho works are collected today.

Watt and Boulton quickly became friends. Boulton too, was inventive, curious, and something of a Natural Philosopher. Watt realized that Soho was the place that could provide the skill and craftsmanship he needed to build his engine. Boulton grasped the growing industrial need for power more clearly than his contemporaries. The two men were anxious to form a partnership to build the new patent engine. However Watt was already bound by his agreement with Roebuck. On his return to Scotland Watt urged Roebuck to bring Boulton into the enterprise. Eventually, after some negotiations, Roebuck offered Boulton a licence to build engines for the counties of Warwickshire, Staffordshire and Derbyshire. Boulton rejected this and in 1769 wrote a letter to Watt which shows his tremendous insight. It has been called one of the greatest documents of the Industrial Revolution. '... I was excited by two motives to offer you my assistance which were love of you and love of a money-getting ingenious project ... my idea was to settle a Manufactory near to my own ... and from which Manufactory We would serve the World with Engines of all sizes ... we ... could execute the invention 20 Per Cent cheaper than it would be otherwise executed, and with as great a difference in accuracy as there is between the Blacksmith and the mathematical instrument maker: it would not be worth my while to make for three Countys only, but I find it very well worth while to make for all the World.'

Above   The Soho manufactory. In the 18th century it was one of the most modern developments of its kind anywhere in Europe.

Below Apart from his designs for
mechanical engines, Watt was also an
able surveyor and civil engineer. His
designs included this bridge at Hamilton
in Scotland.

5 *The Move to Birmingham*

For a time Watt returned to civil engineering. He surveyed a water supply project for his native Greenock and more canals. On some projects he supervised construction as well as carrying out the initial survey. He also made improvements to surveying instruments and invented a telescopic rangefinder for his work.

Dr Roebuck's business affairs had not been prospering – he had conceived ambitious schemes for the manufacture of sulphuric acid, alkali and iron. The flooding of his coal mines near Bo'ness put a stop to this and led to bankruptcy in 1773. After some negotiations Matthew Boulton took over the major interest in the steam engine patent – Watt remarking that none of Roebuck's other creditors valued the engine 'at a farthing'. Watt dismantled the experimental engine, by then rusting in its shed at Kinniel, and sent it by sea to London for carriage to Birmingham. However he continued survey work and a few months later was in the Great Glen, still today a very desolate place. He wrote that for three days, water had kept him 'as wet as water can make me' and he could scarcely preserve his notes. It was here that news reached him that his wife was dangerously ill. He hurried home to find that she had died in childbirth.

Watt was left with two children. He entered a period of depression and complained bitterly to his friends about Scotland, its weather, and his poor finances. He particularly hated 'bustling and bargaining', but his engineering engagements constantly required it. He commented, 'the

engineering business is not a vigorous plant here. We are in general very poorly paid, this last year my whole gains do not exceed £200'. However Watt was in demand and knew he could earn a living in this way. If his wife had lived it seems unlikely that he would have entered a new and risky business developing his engine. He had, though, become 'heartsick of this cursed country' and in 1774 left for Birmingham. Matthew Boulton made his former home available to Watt on arrival. Boulton's friendship too must have been a big influence in helping the cautious Watt to change the course of his life.

Watt re-started experiments at Soho. As before piston leakage was a major problem. However the practical Watt engine became a reality with a new, iron cylinder made by Boulton's friend, John Wilkinson, ironmaster at Bersham, near Wrexham. Wilkinson had recently invented a new system for boring cannons. Older boring tools centred themselves in the cylindrical part of the bore already cut. The tool was likely to wander, especially if the cast metal varied in hardness. Wilkinson devised a rigid bar, mounted in bearings at each end. As it turned inside the rough-cast cylinder, a cutter was passed along the bar. This could produce a circular and truly cylindrical bore.

Watt had never previously had the benefit of an accurate cylinder to work with. He commented grudgingly, 'Ye cylinder is not perfect ... yet there doth not appear any gross error'. The experiments were now successful enough for Boulton to prevail upon the doubting Watt to design two full-size engines. The first was for pumping air to blast furnaces at John Wilkinson's New Willey Ironworks. The next was a 50 inch bore engine and can be regarded as the first real

Left 'Fairbottom Bobs' – an atmospheric engine of the Newcomen type. It was originally sited near Ashton-under-Lyne. It has now been restored and is erected at the Ford Museum in America.

production example. It was erected at the Bloomfield Colliery near Dudley. The starting ceremony was held in March 1776 and a Birmingham paper reported that visitors and workmen adjourned to dinner. The hospitality of the proprietors must have been generous, for the engine was later named 'The Parliament Engine amidst the Acclamations of a number of joyous and ingenious Workmen.'

There were inevitably teething troubles and Watt reproached Boulton with starting too soon. However the new engines quickly proved themselves for power and economy. A Watt engine would do the same work as a bigger Newcomen engine, using only one third of the coal. Only three years after Watt arrived in Birmingham several of his engines were at work and orders had been received from as far away as Scotland and Cornwall.

In 1777 a Frenchman, J. C. Périer, visited the ironmaster Wilkinson and tried to persuade him to make a Watt engine for export to France (where Boulton and Watt had no patent protection). This shows that industrial espionage is nothing new. On another occasion, an engineering drawing of the engine was found to be missing after a deputation of Cornish mine-captains had visited Soho.

Cornwall offered tremendous opportunities for the supply of Watt engines. The coal used for pumping the tin and copper mines free of water was imported at great expense from outside the county. The partners received many enquiries and soon Watt journeyed there to supervise the building of the first engines (it took him four days travelling to reach Truro from Birmingham). Local engineers had long experience of Newcomen engines. Many reckoned to know all

Above One of Boulton and Watt's pumping engines. The massive scale of the engine can be seen by comparison with the figure in the foreground.

there was to know about pumping and were suspicious of Watt. One of these was Richard Trevithick (the man who had taken the engine drawing at Soho 'under a misapprehension'). Watt called him 'impudent, ignorant and overbearing'. On another occasion Watt showed his opinion of local skills when he wrote that, 'Grease cannot be afforded particularly here where the Engine men eat it'. However Watt engines soon proved their value and 'the Engine men' were won over.

Watt spent nearly the whole of 1778 supervising engine building in different parts of the county. So much time was spent there that Boulton bought a house for the partners to use. The business was becoming extremely profitable but Watt became obsessed with money matters. Boulton had mortgaged some engine royalty payments with bankers to raise capital. This terrified Watt who foresaw ruin and continually wrote anguished letters to his partner. Boulton seems to have treated these almost neurotic worries with endless patience. On one occasion he suggested that Watt should take comfort in prayer, morning and evening 'after the manner of your countrymen'.

The royalties, about which Watt worried so much, were in fact, the only way in which the firm received payment. When a mine owner ordered an engine, Boulton and Watt acted rather like consulting engineers. They supplied the drawings and expertise. However the mine owner would employ local labour to build the structure of the engine and would pay suppliers like Wilkinson directly for cylinders and ironwork. Only small parts requiring particular accuracy in manufacture were produced at Soho. The payment to the firm was based on a proportion of fuel costs saved compared to a 'common' engine – originally one third of the saving for twenty-five

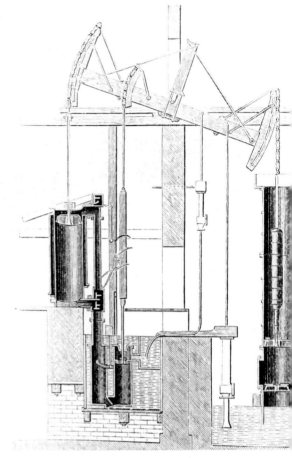

Above Watt's single-acting non-rotative pumping engine. This type used only the downstroke. Later engines were developed to use both the up and the downstroke.

Above Cosgarne House was bought by the partners as a base for their Cornish operations.

years. In Cornwall, Watt engines were often used to pull water from deeper levels than Newcomen engines could pump. No direct comparison was then possible. Watt calculated the payments by comparison with a hypothetical Newcomen engine according to complex formulae of his own devising.

The mine captains mistrusted these calculations, which they could not understand. There was even the threat of an appeal to Parliament against the patent, on the grounds of monopoly. This came to nothing, but it inspired Watt to write, 'They say it is inconvenient … to be burdened with the payment of engine dues; just as it is inconvenient for the person who wishes to get at my purse that I should keep my breeches pocket buttoned. It is doubtless also very inconvenient for the man who wishes to get a slice of the squire's land that there should be a law. … Yet the squire's land has not been so much of his own making as the condensing engine has been of mine. He has only passively inherited his property, while the invention has been the product of my own active labour and of God knows how much anguish of mind …'

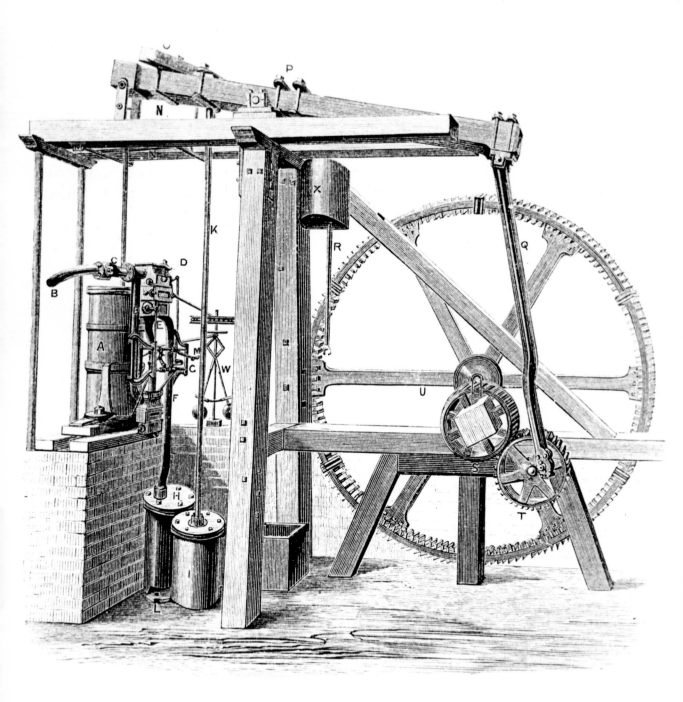

6 Boulton and Watt – Partners

'I sell here Sir, what all the world desires to have – POWER' Matthew Boulton said in 1776 to James Boswell, biographer of the famous Dr Johnson.

Five years later Boulton was still pressing this theme. He wrote to Watt in Cornwall where he was busy with the profitable mine-pumping business 'The people in London, Manchester and Birmingham are *steam mill mad*'. He urged Watt to apply his mind to producing rotary motion from the engine. The tremendous upsurge in all types of manufacturing during the eighteenth century was mainly powered by water. However sites with ample water to drive the wheels were becoming harder to find. In addition, new manufacturers wanted to build in the growing industrial centres – close to raw materials, and other component manufacturers and labour. Steam power would allow industry the choice of location. For these reasons Boulton urged that the next field for the engine was in driving factory machinery. He added in his letter, 'There is no other Cornwall to be found'.

Watt quickly started work at Soho on an engine suitable for mill use. It was of the familiar beam type. Instead of a pump, though, one end of the beam was to drive a flywheel and shaft. The obvious way to do this was to use a connecting rod and crank. However the crank formed part of a patent recently taken out by another steam engine builder. Almost certainly the crank patent could have been challenged. (The principle had been

Left Watt's original beam engine had to be adapted to enable it to provide steam power for the mills.

49

well-known long before, and was in common use, for example in treadle lathes.) Possibly out of pride Watt resolved to do without it and devised his well-known sun and planet (or epicyclic) gear to couple the beam to the flywheel. Although the linkage looks complex, it worked extremely well and the firm continued to use it even after the patent had expired and the crank was freely available.

Watt introduced another change in the engine at this time. The working cylinder was made 'double acting'. By admitting and condensing steam alternately on either side of the piston, the engine gave power on both up and down strokes. This produced more even impulses to drive the flywheel. More importantly, it virtually doubled the power available from the engine.

On the earlier pumping engines, where the down-stroke alone had done the work, the piston was connected to the beam by a chain. Now Watt had to devise a linkage that would both pull and push. A problem was that the piston rod had to move up and down in a straight line through the steam-tight gland on top of the cylinder. Later steam engines used accurately machined parallel guides for this. Watt considered these but rejected them as too difficult and expensive to make with available tools. He spent some time puzzling over this problem. He realized that the piston rod could be guided in a straight line by simple links if these were arranged geometrically in certain ways. His experience with instruments must have helped him, for his final design, called the 'parallel motion', has similarities with the pantograph – a device for copying or enlarging drawings.

The parallel motion has been rightly called 'elegant', not because it looks beautiful – it doesn't. Like most good engineering it used

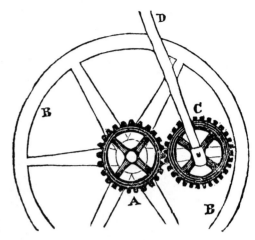

Above A detailed view of Watt's famous 'sun-and-planet' gear which he designed to couple the beam of his engine to the flywheel.

Below Watt's 'parallel motion' design is the perfect example of his ability to combine the eye of the instrument maker with the skill of the engineer.

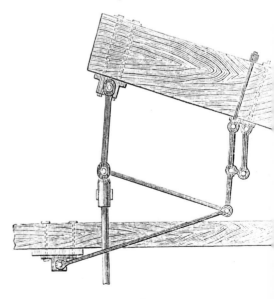

Above Boulton and Watt's rotative engine built in 1788, it is now in the Science Museum in London. This particular engine was called the 'Lap' engine and it was used to drive polishing machinery in the Soho manufactory.

ingenuity to produce an effective solution simply, to a difficult problem. Watt said of it 'I am more proud of the parallel motion than of any other mechanical invention I have ever made.' Until quite recently this 'Watt's linkage' was a common feature of the suspension of racing and competition cars, used to locate the rear axle.

The new power source was a great success. An early application was at the Albion Mill built in London near Blackfriars Bridge. The new mill became a famous showpiece – so much so that Watt wrote complaining about the time wasted by visitors. It also attracted the hatred of rival millers. They rumoured that the owners were attempting to establish a monopoly and force up the price of

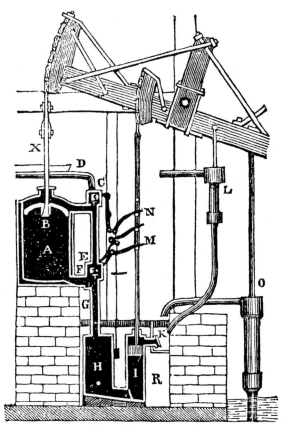

Left A diagramatic view showing the working of a Boulton and Watt engine.

flour. There was much rejoicing when the mill was burnt down in 1791. (Arson was suspected, but the fire seems to have been an accident.) The Albion Mill had in fact shown that the new industrial techniques could reduce the real cost of necessities. During its brief career it had been able to cut the cost of milling corn by over 30 per cent.

The introduction of the steam engine to mill work raised new problems of working out how large an engine was required and charging for it. The calculations had been relatively simple for an engine pumping water. Now Watt needed to devise a unit to measure the rate of working of his engines. Many mills were then powered by horses – the horse walked in a circular track or 'gin' pushing a yoke. This drove the mill machinery through gears. Watt's notebook for 1782 has an

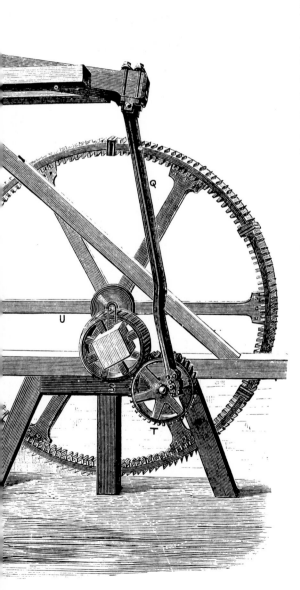

Below 'Old Bess' – the first steam engine that was designed by Watt to produce rotary motion.

entry: 'Mr Worthington of Manchester wants a mill to grind and rasp logwood. ... The power for all which is computed to be about that of 12 horses. Mr Wriggley, his millwright, says a mill horse walks in 24 feet diameter and makes 2½ turns p. minute. ...'

From these figures (and assuming a pull by the horse of 180 pounds) Watt worked out that one 'horsepower' was equivalent to raising 32,400 pounds one foot per minute. The next year he rounded the figure up to 33,000 – probably to aid calculation and noted 'Each horse = 33,000 lb. 1 foot high per minute'. This unit became the accepted one to measure the rate of doing work in all types of rotary power sources up to the present day.

The idea of comparing the power of machinery to that of horses was not new. Watt, however, with his methodical mind, was the first to work out a figure and apply it consistently. Boulton and Watt charged their customers a fixed royalty on their engines of £5 per horsepower per year (or £6 in London).

In other ways too Watt's engineering practices were novel. He was very reluctant to delegate or trust other people's abilities. Thus a tremendous amount of design work fell on his shoulders. In addition, virtually every engine supplied by Soho was different and tailored to the customer's requirements. To aid his calculations, Watt adopted the slide rule, apparently the first engineer to do so. (Until the arrival of the modern electronic calculator, the slide rule was the virtual 'hallmark' of the mechanical engineer.)

Boulton and Watt kept up an enormous business correspondence – particularly since one or other of the partners was often away from Soho supervising engine building and business arrangements. In those days there was no method of copying letters except by writing them again by hand. The labour of keeping track of correspondence must have been tremendous. Watt set about devising a special ink formula. A letter written with his ink would print off a copy when pressed against damp absorbent paper. (Since the copy was a 'mirror image' of the original, it was read from the reverse side through specially thin copy paper.) Watt produced two forms of press for this copying technique and the partners started to sell presses and the special ink commercially. Boulton showed it off in London and sold 150 presses in the first year (despite fears from bankers that they would simplify forgery). The copy press became the standard way of copying correspondence until the development of typewriters and carbon paper. Watt also used the copy press extensively at Soho for his engineering drawings – again, the first engineer to do this.

As the eighteenth century drew to a close, Watt's steam engine patents were increasingly attacked by other builders. In the patents, his discoveries were described in very wide terms and it was virtually impossible to build any improved form of steam engine without enfringing them. However Boulton and Watt did not have the design and manufacturing capacity to satisfy all the demand for their engines. Neither would they licence any other manufacturers. For these reasons, various 'pirate' builders erected 'Watt' type engines. When the partners detected this they would demand their normal royalites from the mine or mill owner, and attempt to stop the builder with a legal injunction. Of course anyone could still build an obsolete Newcomen type engine. It is remarkable that as many of these 'common' engines, with their higher fuel

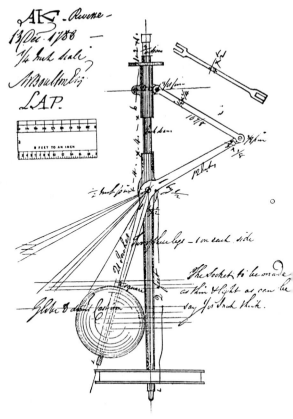

Above A page from Watt's notebook showing his designs for the centrifugal governor.

consumption were erected in the term of Watt's
patents as in the sixty odd years before the Watt
engine became available.

Many of the challengers were from Cornwall,
where long experience of mine pumping had
produced a breed of ingenious and skilful
engineers. In fact some of them, Richard
Trevithick and the Hornblower brothers, Jonathan
and Jabez, were the sons of men Watt had dealt
with when he first went to Cornwall in 1777. Watt
was indignant about virtually all professed engine
builders, saying of Jabez Hornblower '*he calls himself*
an engineer' and using names for them in his
letters like 'Trumpeters, Wasp and Wolf' (the
Hornblowers, Matthew Wasborough and Arthur
Woolf). Jonathan Hornblower in 1781 introduced
a compound engine with two cylinders in which
the steam acted in succession. The intention was to
obtain better efficiency by allowing the steam to
do work by expansion. Watt had in fact patented
the idea of using the expansive power of steam. He

also regarded the second cylinder as a disguised form of his own invention, the separate condenser. Jabez migrated to London where he practised as an engine builder, also using the separate condenser.

Watt and Boulton seem to have been reluctant at first to take legal action against the 'pirates'. Watt, in particular, feared that the courts would not uphold the validity of his patents (he noted that this had happened to Richard Arkwright, the first man to successfully mechanize cotton spinning). Watt remarked, 'I abhor lawsuits and reckon on a cause half lost that is litigated'.

Eventually the pressure of competition from infringers forced the partners to act, Watt claiming, 'The rascals seem to be going on as if the patents were their own ... since the fear of God has no effect on them we must try what the fear of the devil will do'.

In fact a great deal of Watt's time after 1790 was spent in litigation. In 1792 he attended the House of Commons to oppose the extension of Jonathan Hornblower's compound engine patent. Watt's comments on the House were as dry as on his first visit there twenty-five years earlier, from Scotland, 'The people here are in general as absurd as ever ...'

Eventually the partners were successful in their legal actions against the pirates and the validity of the patent was upheld. Boulton was an expert at 'lobbying' and for the final hearing leading scientists of the day assembled to testify for Watt. These included Herschel, the astronomer, and Watt's old friend, Professor Robison. Robison undertook the arduous journey from Edinburgh to support Watt's claim to have been the first to devise the separate condenser. Robison was ill and in constant pain at the time – his journey was a

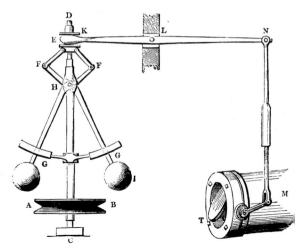

Above Watt's steam governor: as the spindle (CD) rotates, the balls swing out within the grooves (GG), pulling down the collar (E), and eventually closing the damper (T).

powerful testimony to the loyalty of Watt's friends. Dr Black is said to have been moved to tears when he heard the result on Robison's return, saying, 'It's very foolish, but I can't help it when I hear of anything good to Jamie Watt'.

At Soho there was great rejoicing and cannons were fired to celebrate. Although the patent now only had a year to run the decision was worth a considerable sum in royalties – many engine owners had been withholding payment until they saw the outcome of the legal action.

It has been claimed that the patent protection given to Watt was too wide. On the other hand, nearly all the infringers had gained their experience by building or maintaining Boulton and Watt engines. In a sense, the partners did end up teaching the engineers of their day 'fire-engine making'.

Another criticism of the Watt patents is that they held up technical progress (particularly the use of high pressure steam) after Watt perfected his rotative double-acting engine. However, the apostle of high pressure steam was Richard Trevithick. While the partners fought the 'Trumpeters' and the other 'Imps of Satan' through the courts, Trevithick went largely unchallenged and built three high pressure winding engines in Cornwall. He showed that high pressure steam brought greater economy and made possible smaller, even portable steam engines. Trevithick went on to build steam locomotives – both demonstration and working examples. This would have been impossible with the huge beams and massive ironwork of low pressure condensing engines. After the expiry of Watt's patents there was a general trend towards the use of steam at higher working pressures. Watt himself, however, never approved of the practice.

Below James Watt was always subject to fits of depression. Even in later years when he was a respected and successful figure, the thought of failure was a constant dread.

7 The Respected Engineer

From the time that Watt moved to Birmingham he was engaged almost continually in steam engineering and his other mechanical work. He had few interests beyond this. We know too that he had little time for activities like music or painting, feeling that these were a distraction from the serious business of life. Watt's favourite relaxation was conversation with people who shared his keen interest in scientific and technical progress.

He was particularly fortunate that the Birmingham area offered just such company as this. From the 1750's a group developed with a powerful interest in all the latest developments, including electricity, chemistry and industrial techniques (then often called 'useful arts'). The group tended to meet on evenings near full moon, to benefit from its light for travelling home. Thus they came to call themselves the Lunar Society. Matthew Boulton was one of the most important members. His personality drew the group together. Other early members were Josiah Wedgwood, the famous potter and Dr Erasmus Darwin (his grandson Charles was to write *The Origin of Species*). Long before he left Scotland, Watt wrote frequent letters to members – particularly Dr William Small. Small and other members gave Watt much encouragement at times when he grew doubtful about his 'steam improvements'.

The 'Lunatics' as they sometimes called themselves fell on new books and papers with relish. In addition they performed their own

Left The members of the Lunar Society included some of the foremost names in 18th century industry and science – Erasmus Darwin, Josiah Wedgwood and Joseph Priestley.

Above During riots in Birmingham a mob attacked and burnt down the house of Doctor Joseph Priestley. He later left England and went on to live in America.

experiments on chemistry, electricity and natural history. Much of this was motivated by genuine scientific curiosity. Other experiments were practical, as in Wedgwood's trials to improve pottery clays and glazes. However these practical experiments were not purely commercial – there was an idealistic belief abroad that improvements in manufacturing would benefit mankind in general. Wedgwood and Boulton illustrate this movement well. They desired to bring tasteful and useful products to people who could not previously afford them, and to lower the cost of necessities. Broadly they succeeded. The members kept in remarkably close touch with each other, by letters and meetings. They also provided apparatus for each other and exchanged results of

Right Heathfield House near Birmingham was James Watt's home for many years. It was here in 1819 that he died.

their experiments.

In 1781 Joseph Priestley, by then England's foremost chemist, came to live in Birmingham and joined the circle. Probably he had been drawn there by contacts with various 'lunatics' – particularly Matthew Boulton. The manufacturers must have been particularly pleased to have Priestley in the area and assisted him financially with his researches. At one time Watt became seriously alarmed that the Hornblowers had devised a new type of engine, using, perhaps, hot smoke from burning wood or straw rather than steam. He wondered if it were possible and Boulton consulted Priestley, reporting to the anguished Watt: 'I have raised a confidence in Dr Priestley who has seriously promised me to keep the secret and to give me all ... he can discover in that line. The Dr is under some obligation to me that you know not of and I think will not betray me. I am sure he will not.'

Priestley was a man of advanced liberal views. He supported the French Revolution and American Independence and was also a Minister of the non-conformist Unitarian church. In 1791 riots broke out in support of 'Church and King'. The mob destroyed dissenting meeting houses and burnt down Priestley's house, library, and laboratory. Many of the Lunar Society shared his views and feared the mob would turn on them. The rioting continued for three days while the Soho factory was guarded by loyal employees with muskets. Priestley left for London and then America. From this time the Lunar Society gradually declined.

One chemical interest of the Society involved Watt in an unusual project. Some researchers wondered whether the newly discovered gases could have medically useful properties.

Wedgwood contributed £1,000 towards a 'Pneumatic Medical Institute' opened in Bristol by Thomas Beddoes. Watt designed the apparatus to make the gases and it was produced by the firm of Boulton and Watt. Watt's son by his second marriage, Gregory, was suffering from consumption and eventually died. This probably led to his father's interest. Unfortunately, treatment by inhaling various gases such as oxygen, carbon monoxide and dioxide did not prove effective. The hopes of 'pneumatic medicine' proved to be too optimistic. However the Institute marked the beginning of an attempt to give medical treatment a chemical basis. Humphrey Davy, Beddoes' assistant, became a great scientist. While at the Institute he noted the effects of nitrous oxide (laughing gas). This was treated as a curiosity but later became important in early anaesthesia.

Below The famous garret workshop at Heathfield where in his final years Watt occupied his mind with various experiments including the designing of a copy-mill.

Above After a life time of hard work and numerous disappointments, James Watt was a respected figure – the first true 'mechanical engineer'.

Watt and Boulton gradually handed over the engine business to their sons (also called James and Matthew). Boulton senior never really retired, but turned to various projects, including the minting of coins by steam power. This enterprise was successful and brought him further fame. Watt, on the other hand was content to enjoy his retirement. In his old age he became a figure of great dignity. He travelled widely, with his wife, and was treated everywhere with respect – even reverence. Sir Walter Scott met him in Edinburgh in 1805. The novelist recalled that Watt '... was not only one of the most generally well-informed, but one of the kindest of human beings. ... The alert, kind, benevolent old man had his attention alive to everyone's question, his information at everyone's command.'

When at home Watt experimented with his sculpture copying machine, remarking 'without a hobby-horse, what is life'. He feared that he might lose his mental powers with age and set himself to re-learn German as an exercise – he had learnt it many years before, in Glasgow, to help him with his mechanical enquiries. In fact he remained alert to the end and died after a short illness at the age of 83.

Watt outlived most of his old friends. Boulton had died ten years earlier. Doctor Joseph Black and Professor Robison too had died, though Robison survived Black to collect and publish his friend's famous lectures which had influenced many students. Watt never forgot his debt to these early friends. He wrote on Black's death; '... he taught me to reason and to experiment in natural philosophy, and was always a true friend and adviser ... to him I owe in great measure my being what I am'.

Date Chart

1736	James Watt born on the 19th of January at Greenock.
1754	Starts to train in Glasgow as a scientific instrument maker.
1755	Spends year in London learning scientific instrument making.
1757	Opens shop in the College of Glasgow.
1763	Repairs and studies a model Newcomen engine belonging to the college.
1764	Marries his first wife, Margeret Miller.
1765	Devises the separate condenser steam engine.
1767	Takes up land surveying for civil engineering projects.
1768	Enters partnership with Dr Roebuck to develop the condensing engine. Visits London to take out patent. Returns via Birmingham and meets Matthew Boulton.
1773	Bankruptcy of Roebuck. Watt's first wife dies.
1774	Boulton takes over Roebuck's share in the engine patent. Watt moves to Birmingham.
1775	Boulton and Watt successfully petition Parliament for a 25 year extension of the patent.
1776	Watt engines built at New Willey Ironworks and Bloomfield Colliery. Watt visits Scotland and marries Ann MacGregor.

1777	Watt travels to Cornwall and builds engines there.
1780	Patents letter copying process.
1781	Jabez Hornblower patents compound steam engine. Watt patents substitutes for the crank.
1782	Watt patents the double-acting rotative engine.
1783	First rotative engine built.
1785	Watt elected a fellow of the Royal Society.
1791	'Church and King' riots in Birmingham.
1793	Boulton and Watt begin legal action against patent infringers.
1797	Watt studies medical effect of gases.
1799	Validity of the patents upheld.
1804	Starts work on machines to copy sculpture.
1804	Death of Matthew Boulton.
1819	Watt dies on August 25th at Heathfield.

Glossary

CONDENSATION The change of vapour into liquid.

CONDUCTION Transmission, usually of heat through a solid body.

CRANK Device for converting up-and-down (reciprocating) motion into rotary motion – like the starting handle on old cars.

CYLINDER The working cylinder in a steam engine is the chamber in which steam acts on the piston.

ENGINE MAN The man in charge of a mine pumping engine.

FLYWHEEL Rotating heavy wheel that smooths out the uneven power strokes in a piston engine.

LUNAR SOCIETY Small group of men who exchanged ideas and views. Included Watt, Wedgwood and Erasmus Darwin.

MONOPOLY To have the sole right to make something.

NATURAL PHILOSOPHY Broadly equivalent to what is now called Physics.

PATENT The exclusive right to make, use or sell an invention for a number of years.

PIRATE A person or persons who infringe a patent without permission.

PISTON The piston is a form of plunger sliding up and down in the steam cylinder. In Watt's engines it is acted on by steam pressure and it must be a good fit to the bore.

PRESSURE Defined as force per unit area. Experimenters in the 17th century realized that gases (including the atmosphere of the earth) exerted pressure on solid objects.

RADIATION The act of sending out rays of heat.

ROYALTY Payment for the use of someone's patent or invention.

TRADE GUILD Organization for protecting the interests of particular groups of craftsmen – e.g. masons, clockmakers.

VACUUM A perfect vacuum is unobtainable. In general the term means a space containing air at a very low pressure.

Further Reading

James Watt, L. T. C. Rolt, is a readable and comprehensive account (B. T. Batsford 1962)

Other useful sources include:

James Watt and the Separate Condenser, R. J. Law, a Science Museum Monograph. H.M.S.O. 1969

The Steam Engine, R. J. Law, a Science Museum Booklet. H.M.S.O. 1970

The Garret Workshop of James Watt, H. W. Dickenson, a Science Museum Reprint. H.M.S.O. 1970

James Watt and the Steam Revolution, Eric Robinson and A. B. Musson. Adams & Dart 1969

The Classic Biography of James Watt is *Lives of Boulton and Watt*, Samuel Smiles. John Murray 1865

Index

Picture Acknowledgements

Mary Evans 12, 17, 20, 57, 58, 61, 63; Mansell Collection front cover, 33, 42; National Maritime Museum 14; Ann Ronan Picture Library 37, 39, 50 (top), 52, 53, 55, 56, 62; Royal Scottish Museum 10; Science Museum, London Frontispiece, 10 (top), 11, 23, 24, 26, 29, 30, 45, 51; All other pictures from the Wayland Picture Library.